C++

C++ and Hacking for dummies. A smart way to learn C plus plus and beginners guide to computer hacking (C++ programming, C++ for beginners, hacking, how to hack, hacking exposed, hacking system)

C++
C++ for Beginners, C++ in 24 Hours, Learn C++ fast!

STANLEY HOFFMAN

Copyright © 2015 Stanley Hoffman

All rights reserved.

ISBN: 1533279691

ISBN-13: 978-1533279699

CONTENTS

Introduction i
Chapter One – Your First Program 1
Chapter Two – Variables 4
    Using Variables 5
Chapter Three - Expanding Your Program 8
    iostream Library 9
Chapter Four – Operators 11
Chapter Five– Conditionals 13
Chapter Six– Loops 16
Chapter Seven– Arrays 20
Chapter Eight– Functions 23
Chapter Nine– Pointers 26
Chapter Ten - Dynamic Memory 29
Chapter Eleven - Classes and Objects 32
Conclusion 36

I think next books will also be interesting for you:

Python

Computer Hacking

MATT BENTON

COMPUTER HACKING

THE ESSENTIAL HACKING GUIDE FOR BEGINNERS

Introduction

C++ is an up and coming programming language that's based on objects and compiling. If you're someone who's never programmed before, you might be confused by the words object and compiling.

A program that utilized compiling is read by the computer's CPU rather than by being interpreted by a program installed on the computer to run that program. Therefore, the program speaks more to the computer's processing system rather than a program.

You may already know this, but computers are not able to read letters and have to read binary code, so the code has to be translated into a computer readable one. This process is known as compiling. Now, object-oriented code is just how the code is structured. You'll understand that tidbit a little better in the following chapters.

So what does that all mean and why do you need to know it? As a programmer using C++, you're going to need more than just the average text editor. There only a few programming languages where that's all you'll need, and they're not really programming languages. One of the things you are going to need is a compiler.

Another thing you're going to want is an Integrated Development Environment, also known as an IDE. This is a programmer-oriented text editor that has a syntax highlighting and is compiler integrated. For Windows and Linux, you can use a few different options.

You don't really need an IDE because a basic text editor will do, but something like Notepad is going to make things a little more difficult. Word processors are also a bad choice because they have autocorrect with spelling and grammar, which will mess up your code if you're not careful.

To begin with C++, you should understand that it has a rigid syntax. This means the code is case sensitive, so typing 'Hello' is going to be different than typing 'hello'. In addition, you should know that the programming language doesn't care much about white space, so don't worry about it.

So that's about all you need to know about C++ when you first begin. Let's get into some actual coding to get you started!.

Chapter One – Your First Program

Due to this being a short tutorial on how to get started with C++, we're going to start you out with a program in the first chapter. Your first program is going to be very simple. All you're going to do is write a single phrase on the screen, but it's not as simple as writing it in a Word document. First, let's take a look at the code and then we'll go further into what it means.

```
1          #include <iostream>
2  using namespace std;
3
4  int main()
5  {
6      cout << "Hello, World!";
7
8    return 0;
9  }
```

The line is going to tell the compiler it has to link to the isostream library to the program. A library is a repository where snippets of code are stored, known as functions. The library also stores other things, like variables and sometimes operators.

The isostream library has code that will let you send characters into the text stream, which is then fed onto the screen. Think of it like a river. You're the river's source and the screen is the sea. Whatever is put into the river from the source is going to go onto the screen.

The following line, line two, lets the compiler known that you want to use the namespace known as std, which is a standard namespace. Namespaces are an area where the code is stored. This is useful if you have two pieces of code that have the same name. Normally, pieces of code are not able to share the same name, but if one of is in a separate namespace than the other, then they can have the same name.

The third line of code is the beginning of the code. It's the main block, as the name implies. Everything inside the braces {} will be translated by the compiler. You'll need that line in every program because nothing will happen without it.

Now comes the printing of the phrase, 'Hello World!' To do this, you use code that's stored inside the isostream library. This part of the code is known as the statement or cout, which sends information to the stream. The << that comes after the coutis an operator. This means 'output', so with the coutit all translates to 'output this stream'.

The following stream is 'Hello, World!' and is known as a string. Strings are a sequence of characters and are shown by enclosing them in double quotation marks. This ends with a semi-colon and signifies the end of the statement. If the semicolon is not used, then the program is not going to compile.

The last line, not the closing brace, is very simple but also very important. It's placed at the end of main and tells the computer that the main block has been finished. It will return a value of 0 to the CPU. A non-zero value will create an error, so it's a necessary line.

Now that you have the source code of the program, you have to compile i. If the program closes once you run it, then try running it from the cmd/terminal prompt instead.

Now that you know how to create your first program and what it looks like let's take a look at variables!

Chapter Two – Variables

In the last chapter, variables were mentioned and it was said that they would be discussed. Now is the time to discuss them. A variable is something that stores a value. Imagine that every variable is a box and every box is labeled with a name and category.

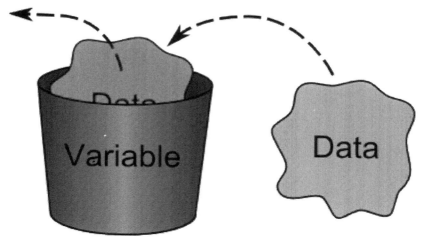

The category is what type of item it's storing and its name is the specific variable it's called. Boxes are able to store many things, but only if that thing fits inside the box. This is the same with variables.

They all have a type, name, and size that govern what they are and are not able to store.

For example, one type of variable is the integer. If you know basic mathematics, then you know that integers are zero, whole numbers, and their negative counterparts. So -5, 4, 0 and -378 are all integers. In addition, integer variables will have a fixed size.

They are only able to hold one number between two others. These are specific to the variable and will not be discussed quite yet, but a default integers range is between -2^{31} = -2, 147, 483, 648 and $2^{31}-1$ = 2, 147, 483, 647. You probably won't need numbers larger than that yet, so we'll save the rest for another time.

Here's a table to see the different types of variables you can use.

C++ Type	Stores
int	Whole numbers (integers)
bool	True or False (0 or 1)
float	Fractional numbers (reals)
double	Float with twice the precision
char	A single ASCII character

There are a few different variable types. You'll probably use them all when you program, but for now you're only going to use a few.

Using Variables

Take a look at the following program:

```
1  #include <iostream>
2  using namespace std;
3
4  int variable;
5
6  int main()
7  {
8
9  variable = 42;
10
11 cout << variable;
12 return 0;
13
14 }
```

Most of this program is going to look familiar, except for the fourth and sixth lines. The first unfamiliar part declares the variable. This is telling the compiler that you're going to use a variable. In declarations, the variable type is going to come first. In the example, it's using the int type. Next is the variable's name, which would be known as var.

It's easy and simple. A variable name can be anything, but be sure it's something relevant and not something that's already reserved and means something in C++ and it doesn't begin with a number or a special character. You should start with a lowercase letter, and end the statement with a semi-colon.

Line nine in the code above is known as an assignment and assigns a value to the variable. The equal sign is the assignment operator, and it is what does the assigning. The variable on the left is given the value on the right, and the = is a binary operator because it has two operands. One to the left and the other to the right. The operand on the right doesn't have to be a value. It can be another variable. The left has to be a variable and you can't change the meaning of the number seven. In the example, the integer variable var is a value of forty-two.

You can see the cout statement doesn't have quotations marks around the variable name. That's because you want to output the value of the var and not display the string 'var'. If you were compiling the program, it would print '42' because that is the value the var has. Try changing it to another integer and look at the output.

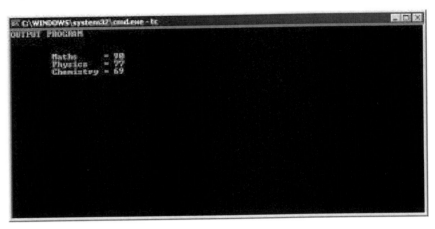

Variables also have prefixes that are added to their declaration that changes how they act. Some will change the size, and others will not allow negative numbers, and others might do something entirely different.

Now that you know about variables let's talk about how to expand the program!

Chapter Three - Expanding Your Program

Now that you know how to use variables and how to put them into the program let's take a look at how to expand the program. All you really know what to do is use the variables and put them into your program, so all you can really do is set the output and values of the variable. That isn't useful when you're creating a real program.

What you need is a way to manipulate the variables and receive input from a user. So let's take a look at the next!

iostream Library

You already know that the iostream library is able to output data, but it can also obtain input data, too. The command is very like the cout statement. The command is cin. There is one different, though. When you use cin, the operator after it's inverted.

Instead of being <<, it's >>. This is confusing to remember in the beginning, but you'll learn it eventually. Note that when you run the program, you'll need to press the enter key after you put in the desired information, like a terminal/cmd prompt.

Take a look at the following program to see the difference between the two.

```
1 cout << "string";
2 cout << var;
3
4 cin >> var;
```

Now you can probably see that it's simple to get the user's input using the iostream library, which produces some more useful programs. iostream is a lot more powerful, though. You can combine many statements into one.

For example, to put out two strings or variables you would normally need two statements, but you can combine many cout statements and produce a much shorter code. All you have to do is just add another operator for the item you're outputting or inputting.

For example:

```
1 cout << "Hello, " << "World!";
```

You can also combine togethercinstatements in the same way you'd receive several parts of the inputin a row. Enter has to be pressed after each one, and you can't mix input and output on a line of code.

Now that you know how to combine code and expand your program let's talk more about operators.

Chapter Four – Operators

You've learned how to output text, declare a variable and obtain input, but you can't do anything with the variables. Sure, you can give them values, but what use is that in the real world? What you really need is a way to manipulate the variables without doing hard work. Luckily, the operators will save you!

You've already investigated two operators that are linked to the iostream library, the input and output or >> and <<.

Operator	Use	Example	Result
+	To add two numbers	i=3+2	5
-	For subtraction	i=3-2	1
*	For multiplication	i=3*2	6
/	For division	i=3/2	1
%	Modular division (Reminder after division)	i=10%3	1

These operators are able to be put into categorical groups, the first one being arithmetic operators. These are -, +, /, and *. You would use them exactly you would use them in math. For example, this piece of code will add two numbers together and display the results.

```
1  #include <iostream>
2  using namespace std;
3
4  int sum = 0;
5
6  int main() {
7
8    sum = 64 + 32;
9    cout << sum;
10
11   return 0;
12 }
```

Now that you know how to use operators let's take a look at conditionals!

Chapter Five– Conditionals

It's great to have a program that does single, linear tasks, like saying 'Hello', but it's rare that you're going to come across for the need of having a program that does linear tasks only. Programs usually use something known as conditions, or statements that can branch the code into different paths depending on whether a condition is true or not. It's like a fork on the road, basically.

C++ conditionals have the form of if statements. For example, if something is true, execute this line of code, if something is false, execute another line of code. Take a look at the following code piece as an example.

1 int a =;

2

3 if (a<2)

4 {

5 cout <<"a is less than 2!/n";

6 }

Lines one through five should be pretty obvious to you by this point, but line three is the one where you may need a little explanation. This is the example of the if statement. It operates by first taking the matter of the integer variable a, and checks to see if it's less than 2. If that's true, then it'll run the code that's in the curly brackets.

If it's not, then it will continue as if the piece of code didn't exist. We know that 1 or the value of a is less than 2, so if you were to run this program, then you'd see 'a is less than 2!' on the screen.

Here there is another operator introduced, the less than one, <. If you're familiar with math, then the use of this sign is pretty self-explanatory. If you're not, then you'll need to learn it. Note that while < looks like <<, they are not even close to the same thing and are not related in programming.

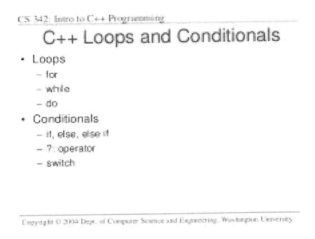

Of course, C++ is a strong language, and so it's able to do more powerful things using these condition if statements.

So in summary, C++ conditionals start with an if statement, and they can have a number of else ifs after them, too. An important thing you should remember is that once one of them has been satisfied, then all the rest is going to be skipped. If you want all of your if statements checked, then do several single if statements one after another.

Chapter Six– Loops

Now you're able to write code that is able to read input from the user, do different things depending on the input, and then show the results, but if what if you want to do the same thing but with some different parameters, several times? It would be boring to write the same code over and over again, and awkward if you don't know how many times you need to run that piece of code when you're writing it. What if it depends on something the user entered?

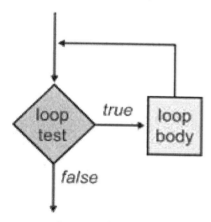

To complete this action, you'll need to use something known as a loop. Loops are a piece of code that are repeated several times, one after another until a condition is met. C++ has three different types of loops. There's the while loop, the for loopand the do while loop. Let's start with the while loop since it's the easiest.

While loops look like if statements, almost.

```
1 int userInput = 0;
2
3 while (userInput !=10)
4 {
5 cin >> userInput;
6 }
```

The use of 'while' is evident when you know what while loops actually do. Here, the code is saying that 'while the value of in the userInput variable does not equal 10, obtain some input from the user'. In addition, remember the ! means 'not' in C++, so the code '!=' translates to not equal.

If the userInput is ten, then the code is not going to be executed and the loop will be skipped. While loops can be seen as being an if statement that jumps back to the information in itself once the code has been run.

Before we keep going, it's important to know that the following piece of code is usable in C++:

```
1 while (1 == 1)
2 {
3 cout << "a";
4 }
```

It actually has its own name. It's an infinite loop so it will loop infinitely. Going through the code, you'd get something like 'Is 1 equal to 1? Yes it is, print an 'a'.' This will repeat over and over again. There are other ways to make infinite loops, but you probably don't want to make one.

The second type of loop is the do while loop. Do while loops look like this:

```
1 int userInput = 0;
2
3 do
4 {
5 cin >> userInput;
6 } while (userInput !=10);
```

While it looks different than the standard while loop you just saw, it's almost the same thing, except there's one difference. The while loop evaluates a condition to see if it's true and will run the code if it is, and the do while loop will execute some code and then check to see if the condition is true. Due to this happening, no matter what you initialized as userInput to the code between the curly-brackets will definitely run.

The third type of loop is the for loop. This is the more complicated out of the three, but it's powerful when you understand it. For now, let's just cover the uses. Take a look at this for loop.

```
1 int i = 0;
2
3 for (i = 2; i < 10; ++i)
4 {
5 cout << i << "/n";
6 }
```

This piece of code tells the program to start by setting the value i to 2 and as long as i is less than 10, then output its value to the screen and increment. You can easily replicate this loop using a while loop, but it uses a lot more line of code and is pretty untidy.

You should also know that the variables used in the first line of the for loop have to be the same variable. You also should remember that the first statement is not a conditional, it's an assignment.

This kind of loop comes in handy when you want to step through various pieces of data, and you'll see this when you read about arrays in the next chapter.

Chapter Seven– Arrays

Arrays are a list of variables that are grouped together because they have a common use. They don't really have to have something in common, but when you put unrelated things together it's not really a good idea. It's like filling your filing cabinet with random papers and never organizing them. What the need for a filing cabinet of you're going to do that?
Here's a declaration of an array of integer variables.

1 int a [5];

You can see that arrays look like normal variables, but they have some square brackets at the ends. When you declare arrays, the brackets are then filled with the size of the array you're looking for. In this case, you're going to store five integer variables.

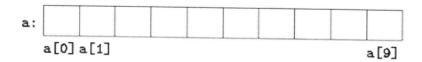

An array can be whatever size you want as long as you have the RAM to spare, but it can hold only one type of variable at once. The variable type can be something you'd normally use.

Accessing a certain variable in the array is simplistic. You put the numbers of variables you want in the square brackets. So for the first variable you want [0], for the second you put [1] and keep going. Take a look at the following example.

```
1  int a [5];
2
3  // Set the first variable in a to 1
4  a [0] = 1;
5
6  // Set the second variable in a to 4
7  a [1] = 4;
8
9  // Output the first variable
10 cout << a [0];
```

One thing you shouldn't do is access a variable that's not in the array that doesn't exist, like the tenth variable in an array that has five places, or doesn't have a variable assigned to it. If it doesn't have a variable assigned to it, you'll get what's known as a garbage value because the value it returns is not going to be useable.

If the variable doesn't exist, you can get a segmentation fault because the program will try to access a segment of memory that it's not allowed to. This can result in the program terminating unexpectedly.

Now that you know how to create arrays let's take a look at creating functions!

Chapter Eight– Functions

Functions are a great way of writing out complex programs without needing to repeat code. For example, let's say you have a complicated part of code that gets some input from the user and parses it. Now let's say you need to run the code several times, but not in a loop. Up until this point, you'd have to write it out every time.

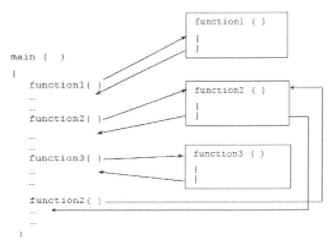

With modern copying and pasting, it's not that big of a deal, but it'll look horrible and having repeated code can be a problem when you go to fix any errors. To remedy this problem, you can create functions, which are pieces of code that you're able to call repeatedly without needing to write it out every time.

Actually, you've already used a function. Main is a function. So let's take a look at it to see how it works.

```
1 int main ( )
2 {
3 cout << "I'm in a function!/n";
4
5 return 0;
6 }
```

Let's examine the first line, the one that has the word main in it. This begins the definition of the function that will take no parameters, known as main, and will give back a value of type int. You already known what an int is, a type of variable, but what does return a value mean?

A function in C++ is like a function in math, you insert a number and some operation is performed on it, and you get an answer. Here, the int is the variable that the output function is going to be. A parameter is the variable that is put into the function, and then the function performs an operation before it returns the results.

Let's take a look at a function that has a main.

```
1 // This is a different way of writing main
2 int main (int argc, char* argv [] )
3 {
4 cout << "I'm in a function!/n";
5
6 return 0;
7 }
```

The parameters are written like a variable definition and is separated by a comma. They are written right after the function name and are closed in by parenthesis. Now there is an integer value known as argc and an array of characters known as argv as the parameter. You can use any type of variable you want as a function parameter, as long as when you use that function you provide a variable that's of the same type.

Now that you know what a function is and does let's take a look at pointers.

Chapter Nine– Pointers

RAM or random access memory is where the computer saves all the variables the program uses, and is basically a long line of numbers. Every number is identified by the address or where it is on that line. These are logical, so the number with the address four comes after the three but before the five.

The exact size of the number at that address is figured out by the computer architecture. Say you have two integers and a character. One might have an address of 1000 while the other has an address of 1004. The character or char might have an address of 1008. So why is there a four byte gap between the variables? Integers are four bytes long, and so they take up four addresses in order to store their value. It doesn't matter if you use one byte and don't use the other three, it will still take up four bytes of memory.

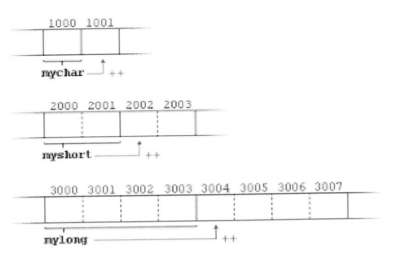

But why do you need to know this? C++ programmers need to know that their compiler does not have fancy addressing for them, translating their variables into addresses in RAM, but sometimes it's good to know the translations on our own. Think about how the arrays work for a moment, they are just long lines of variables whose type is what the programmer tells the compiler.

So you might be thinking this is just like RAM and you'd be right. When you labeled the arrays and the fifth value was actually the number six because all arrays start at zero, you're actually taking the fifth variable after the first variable. So if you say the array is an integer one, with the first integer coming from RAM address 1000, then combine that will what you know about integers being in four bytes, you can see that the nth integer in the array would have this address:

Address of the nth element = 1000 + (4 x (n-1))

or

Address of nth element = Address of 1st element + (variable size x (n-1))

Therefore, the address of the sixth element is 1020. So why is this useful? You can make the array while the program is compiling and running and make it the proper size.

A pointer is a normal variable, but instead of storing the integer, it stores the address of another variable. Here is a pointer.

1 /* This is a pointer */

2 int *foo = 0;

In this pointer, you have a variable that stores the address of the integer. The * is the variable as a pointe rand not a normal variable. You've assigned it to the value of zero. When you assign values to pointers, you're not giving them numbers or characters, you're giving them the address of another variable.

Now that you know what they are, let's move on to dynamic memory in the following chapter.

Chapter Ten - Dynamic Memory

So far you've been able to create variables at compile-time and modify them during run-time; however, you can create them in run-time, too when you use dynamic memory allocation. This method uses allocation and deallocation of RAM as the program is running. Since you're accessing RAM, it's natural to assume that you'll do it with pointers, but dynamic memory in C++ is utilized differently than it is in C.

Take a look at the following block of code.

C++

```
1 int* i = new int;
2
3 *i = 5;
4
5 cout << *i << "at address" << i
6 << "which itself is at address " << &i << std::endl;
7 delete i;
```

Here you've made a pointer to an int and set aside some memory for it. Since i is still the pointer, not a variable, you'll have to set the value pointed to by i to five and not i. This changes the location point by i and orphaned the set aside memory in RAM somewhere. This memory that's been forgotten is known as memory leak. To stop this, be sure to deallocate the memory you set aside in the first place using the delete operator.

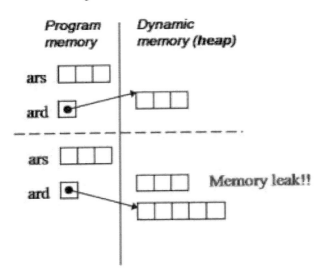

Allocating memory down to single variables doesn't help much, though. If you know the variable was going to exist then you could declare it normally. What this does is create variable size arrays that are not known as compile-time. How would you possibly know it, really? You can't create an array of pointers and tap into them as you need them, that's just a more RAM intensive version of making gigantic array and hoping everything goes well.

Now that you know what dynamic memory is and how to use it let's take a look at classes and objects in the following chapter.

Chapter Eleven - Classes and Objects

This is actually what C++ was made to do, support classes and objects. Without them, C++ is just really C. But what are the classes and objects that have been mentioned before? Classes are like a type of mold and the object is the thing that comes from the mold. These molds are flexible and are able to be used to make whatever type of object you'd like. The molds can be general or they can be specific.

	CLASS
	General Attributes and Behaviors
Name	Apple class
Attributes	* Color * Taste
Behaviors	* Display()

So let's say you want to make a vehicle. This is going to be an abstract tutorial as the code is not actually doing to do anything.

Think of what properties a car needs. They need speed, price, color, manufacturer, and a model. What about if it's four wheel drive or not? And what can you do with it? Can you start the engine and stop?

Let's take a look at the class definition of what the car looks like.

C++

```cpp
class Car
{
private:

bool running;

public:

string manufacturereName;
string modelName
bool fwd;

double topSpeed;
double price;

unsigned int color; // Colors like 0xB9DCED

```

```
18 Car(string manufacturerName, string modelName, double topSpeed, double price, unsigned int color,
bool fwd)
19 {
20 this - >manufacturerName = manufacturerName;
21 this - >modelName = modelName;
21 this - >topSpeed = topSpeed;
22 this - >price = price;
23 this - > color = color;
24 this - >fwd = fwd;
25 this - >running = false;
26 }
27
28 void start ()
29 {
30 this - >running = true;
31 }
32
33 void stop ()
34 {
35 this - >running = false;
36 }
37
38 };
```

So as you can see, classes are defined a lot like structs, and that makes sense because they are both a way to group data. But this class contains something that a struct is not able to, functions.

Conclusion

So by now you know how to create your first, simple program and much more! You know what a variable is, how to expand your program, how to use operators, conditions, loops, arrays, functions, pointers, dynamic memory, and classes and objects to morph your program into a masterpiece! You see, C++ was not as difficult as you may have once though it was.

The next step is to gather the knowledge you've gleaned from this book and go on to read more about the wonderful world of C++ programming. Once you're comfortable, set up a virtual machine to do your coding in and get started! You'll learn a skill that's highly valuable in the computer industry today.

Hacking for dummies
A beginners guide to computer hacking (hacking, how to hack, hacking exposed, hacking system, hacking for dummies, hacking guide, security breach, hacking techniques)

MATT BENTON

CONTENTS

Introduction 40

Chapter 1: Hacking—A Regime of Technological Intelligence 42

Chapter 2: What Do Hackers Do? 49

Chapter 3: Some Major Hacking Ventures—Email Hacking & Windows Hacking 54

Chapter 4: Hacking Web Servers 62

Conclusion 68

Introduction

Hacking has emerged as one of the most attended and discussed topics of the last few years because of the growing number of hacker communities and the extensive effects which they extend towards different personal and corporate computer systems. Even large business tycoons and gigantic empires have been largely affected by this phenomenon of hacking. So whether the victim includes an individual victim or a multiple system of networks and computers, in both cases the aftermath can be drastic. This broader perspective and long-lasting effect of hacking over victims has grabbed the attention of security analysts.

Basically, hacking is successful when there are weaknesses in the security protocols of a system. The hackers who have a proficient knowledge base to detect these loopholes accurately target these kinds of systems. So being a victim of hacking and knowing about it is like following a reverse engineering approach. When you become a victim, you trace the footsteps and come to the point of initiation, which leads the way to the hacker.

This book will provide you the landmarks to follow this reverse engineering. As long as you keep updating your knowledge, you can entail a stronger security system for your computers and networks. It is especially useful for outlining an error-free system of security for your system and network. It is possible you'll start exploring the foundations of knowledge which lie at the root of hacking. In this book I have tried to explore the major dimensions of hacking which can help you better understand the phenomenon.

Chapter 1: Hacking—A Regime of Technological Intelligence

The human intellect is no less than a miracle. The superb and mind blowing use of human intellect can be depicted in different areas and domains of human civilization. This intellect makes the human race the most superior amongst all creatures. Even though wise humans existed even in the earlier days of human civilization, such as during the caveman period, the modern man has exploited his intelligence in a much wider and broader domain.

Although the areas of development and progress are many, the milestone of development pertains to the area of information technology. The spectacular development in this field has literally revolutionized human corporate and household working. The binary digits, though seemingly tiny, have put the world in a new era. Along with the World Wide Web and the internet facilities, today's world is dependent upon the clicks of a mouse.

The corporate world is just an interweaving of clicks and virtual domains; each business and government entity now has hundreds of computers and internet access gadgets, which makes the whole system work. But as there have been both sides to any phenomenon, so does the world of information technology which has also undergone various certain attention-taking trends. One such trend is hacking.

Hacking:

Hacking is an activity of finding and exploiting the weaknesses and loopholes in personal or corporate computer systems. Hacking can be regarded as a consequence of insecure computer systems.

While the losses incurred through a hacker's attack may vary according to the type of data being attacked, the level of knowledge needed for attacking a system is always the same.

The Aims and Intentions of Hacking:

Being a beginner in the domain of hacking, you may think of hacking as some malicious and destructive activity. Although it is true most of the time, the intentions and aims of hackers may vary. A hacker may be indulged in the activity of hacking for:

- Financial gains
- Protest
- Challenge
- Enjoyment
- Destruction
- Professional rivalry
- Assistance

Looking at the aims and intentions of these hacking activities, you may have noticed that hacking is not all about financial gains or destruction. Sometimes induced or known hacking is practiced for constructive purposes.

Hackers and Their Groups

A hacker is a person who has a passion or duty to trifle and trickle computer systems and electronic gadgets. Hackers have extreme knowledge about the workings of these systems, so using their knowledge they can see and locate any kind of weakness in the system. They can even investigate better performance parameters for those systems. Some hackers, by using their knowledge about computer systems, induce these systems to perform functions in a destructive way.

Based on the ultimate intention and result of hacking, we can categorize hackers into two major categories. These categories have been named with inspiration from old Western movies, where the good and pious guy used to wear a white hat, and the opponent was always an evil guy with destructive intentions who always wore a black hat.

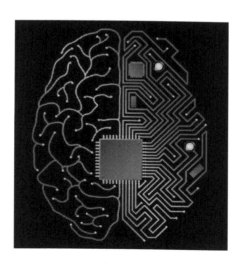

White Hat Hacker

This is the category of good guys. As the name denotes, wearing a white hat is an emblem of peace and constructive activity. White hat hackers are not indulged in any destructive or illegal hacking. They either perform hacking with the permission of the owner, or practice this activity for enhancing their knowledge and study of computer systems. As they are regarded as good champs, they are usually found in the security departments of various governmental and private organizations where their responsibilities pertain to saving the system of the organization from destructive agents.

Black Hat Hackers

This is the category of bad guys. As their name denotes, they are bad with evil purposes. Black hat hackers utilize the knowledge of computer systems and securities for malicious and nasty purposes, mostly for individual profit and gain. You may find a black hat hacker behind bank robberies, stealing credit cards, and defacing websites. They may be operating for individual purposes or they may have a whole network which is involved in these kinds of malicious activities.

Many rival organizations, states, or agencies may even hire these black hats in order to defame their opponents or get access to their systems to attain sensitive information.

The Hierarchy of Hackers

Just as the intentions and purposes of hackers may vary, the levels of expertise also vary tremendously, which creates an hierarchy of hackers. The hierarchy is as follows:

Script Kiddies

These are hackers which you can categorize as "wannabe hackers." They are looked down upon as notorious even in their own community because they do not have any kind of good reputation. They are just like followers. They gain access to one or two major tools which are used in hacking and use them to experiment different hacking practices. So their level of proficiency and knowledge in the hacking community is minimum. They just use hacking code as a script.

Intermediate Hackers

Next up is the intermediate hackers. They possess a moderate level of knowledge pertaining to network systems, computer security, and different codes. So as opposed to the script kiddies, they may know what a script is actually doing to the back of a computer system. They can make an alteration, as opposed to script kiddies who blindly follow. But being at an intermediate level of expertise, these hackers use the already developed codes and procedures of hacking. They tend to select some pre-developed code according to their purpose or intention and use it for their own benefit.

Elite Hackers

Amongst the community of hackers are the elite hackers. Just as there is an elite-class that possesses all the resources and facilities, so do elite hackers possess all the knowledge and expertise at the highest level. But reaching this level is not easy. The hackers struggle and strive to reach to this level. They can write codes and paths to reach and break security barriers. They can alter and modify their hacking activity as and when needed, and therefore they possess the highest level of expertise in the hacking community. They can even hack systems without leaving any footprints of their presence.

Chapter 2: What Do Hackers Do?

The domain of hacking is not a matter of following a few steps. It pertains to a whole set of knowledge and expertise to be applied at different levels of hacking. When pertaining to the knowledge of hacking, it is not always necessary to know about hacking to hack. But this knowledge can help you get away and protect yourself from the activities of these hackers. Sometimes you may need to apply this knowledge to find out the weaknesses and loopholes in your computer system, whether for personal or corporate use.

Hacking is Not Cracking!

There is a very fine line of difference between hacking and cracking. There has been a lot available in the literature of information technology which guides one to know the difference between the two. Whenever there is news about the stealing of data or loss of data in some computer system, people pertain it to hacking.

A hacker is a person who is fond of exploring operating systems and computer gadgets. Programming is one of the most basic knowledge implied by hackers, so you may expect a hacker to have in-depth knowledge of program codes and languages.

They also know about security loopholes within computer systems and the ways to address these deficiencies. The most professional hackers can apply their knowledge to a number of various situations, so their knowledge will help them deal with various situations. They are not just code readers.

Crackers are usually people whose knowledge is lined to the field of entering databases and reading the data. Cracking is mostly done with malicious purposes. So you can consider cracking as a sub-domain of hacking. A hacker may first need to apply cracking knowledge to gain entry into a system of computers and data. Crackers are usually considered to have less knowledge as compared to hackers because they have mastery in only one domain of data retrieval and access.

So when an incident of pirating data occurs, you need to get an in-depth analysis of whether it is hacking or cracking.

The Starting Point of Hacking

Learning hacking knowledge and exercises may come along your way when you become a victim of it or when your organization demands you to learn about it. Many of us may get confused as to what the starting point is.

The basic knowledge of hacking starts with a proficiency in programming. So the initiation point is in the knowledge of Hypertext Markup Language, also known as HTML.

HTML is the basic building block of all website pages and other types of web content. It is governed through the basic knowledge of source code and programming techniques.

The Steps Performed in Hacking

➤ Performing Reconnaissance

Reconnaissance is considered to be the pre-attack stage. It is this stage which entails all the preparation needed to attack the target. It is an organized way to situate, collect, recognize, and trace the information regarding the computer system of the target. The information gathered in this phase will set out the intensity of the attack of the victim. So the better the information gathered, the stronger the attack on the target.

➤ Enumeration and Scanning

This can be regarded as the next step to the pre-attack phase. This phase makes use of the information which has been collected in the first step. This information will be used to read and analyze the typical characteristics of the network systems. Scanning is also performed in this phase. In these procedures all the vulnerable and open ports of entry are located within an operating system. The system vulnerabilities are scanned using different tools and sophisticated programming language.

> **Gaining Access**

This is the stage of real hacking. It is the phase of the exploitation of all vulnerabilities which have been detected in the first two phases; the mode of connection to the target system is developed in this phase. The choice may involve a local access to the operating system, local area network, local access to the Internet, or offline modes. When the hacker gains access to the target system, it is referred to as "owning the system."

Once the access is achieved, the target system is vulnerable to a number of possible reparations and damages.

> **Maintaining Access**

One-time access to the operating system of computers may not be useful for the complete hacking activity, so it is necessary to maintain access. It will not only ensure the complete hacking activity but provide the route for even greater control. So the ground and foundation is strengthened during the access gain phase to allow the hacker to roam about in the system. They fix some points for re-entry and maintenance of access. In this phase, the hacker usually makes the entry exclusive for his own access by using sophisticated tools like root kits, backdoors, and Trojans. This phase mainly involves the use of automated tools and scripts to wash away any signs of hacking. The backdoors are also created using these tools.

> **Clearing Tracks**

Just like how a robber clears away all evidence before leaving, a hacker also clears away all evidence. All tracks and paths are covered in this phase. In this phase the major security tools used involve checking the logs and firewalls. Once this phase is clearly accomplished, there are many chances that the hacking and access will be continued in the same system in the future. It is because if they don't clear their tracks, access is detected in the early phase and the system experts can create even greater controls for hindering their future access.

Chapter 3: Some Major Hacking Ventures— Email Hacking & Windows Hacking

In today's world, email has emerged as one of the most basic ways of connection and communication. Whether you want to greet your friend who is sitting miles apart or you have to email some commercial order for your business activity, the electronic mail is the most economical, trustworthy, and rapid way of communicating.

But because it is being rapidly used, so it is also being exploited by hackers. Basically the sending and receiving procedure of emails is governed through different email servers. All types of email service providers organize and connect to the email server whenever a person signs in to their email account. All digital communication is processed through servers. These servers operate worldwide.

The Security Protocols of an Email System:

Although emails are considered to be the fastest means of communication which are deemed to be reliable and quick, we hear hacking news about emails and their respective accounts. So it infers that in the context of hacking we need to analyze the system of security which is extended by these email service providers and main servers. There is a need to understand that there are vulnerabilities which are exploited by the hackers when they target an email.

Some people think that if email accounts do not contain any important email or personal email, than there is no need to protect these accounts. But email security is not only needed to maintain the security of emails. Hackers can use your email account for malicious or even terrorist activity. So in this case, a strong password will surely help.

Email Spoofing

Email spoofing is a process which involves falsification of a particular email header. In this case the email will appear to have originated from a source or sender other than the real sender. Email spoofing is usually targeted through spam sending, so that when the recipient opens the spam, he may open the way to the hacker to start the spoofing.

Spoofing is sometimes regarded as fake emails and there are multiple ways to send fake emails, even without knowing the password of the email owner.

Sending Fake Emails

There are two major ways for sending fake emails, discussed below:

> ### Using Web Script

The various programming languages which include ASP and PHP can allow a person to use an email send function, in which a fake header can be easily implemented in the form of " From: To: Subject:"

Even you can find email sending scripts on a number of different sites which provides the fake email sending scripts for different purposes.

Most of these websites are anonymous. Some include:

- Deadfake.com
- FakEmailer.info
- FakEmailer.net
- Mail.Anonymizer.name

➤ Open Relay Servers

An Open Mail Relay follows the Simple Mail Transfer Protocol (SMTP) server. Thus the server provides an opportunity to anyone to send emails through some account through its special configuration. It can send emails other than the "Originating" or "To" format.

Now as these servers are available, a hacker can easily use these servers and guide these servers to send emails as and when required. The sending of the email is easier through this method because it does not require any kind of password for opening the specific email address. The hacker will just select a particular email account and the work will be done.

Windows Hacking

Windows possesses a system of security which is based upon the following major components:

- ➤ Local Security Authority (LSA)
- ➤ Security Reference Monitor (SRM)
- ➤ Security Account Manager (SAM)

All of the three components come out as the basic building blocks for securing a Windows account. When pertaining to the LSA it is supposed to be the security subsystem. It validates both the remote as well as the local log of the system. So the basic security policy is translated through LSA.

SAM comprises of passwords and usernames, so the information can be regarded as the one which can be found on the hard disk. SAM information can make use of local registry or that of the active directory. When the SAM database clears the user as part of the server, he or she can then start using the service.

SRM is more like an architectural object which makes use of requests preceded by the user to get access to a number of different objects pertaining to the system.

Learning the Architecture of Windows Security

The passwords for the user account are maintained under a hexadecimal format of SAM, known to be "hashes."

When any password is translated into hashes, it is not possible to turn it back to simple, clear text.

Cracking Windows

All Windows passwords are kept and transmitted in a special encrypted format which is referred to as "hash."Whenever a user starts using a Windows account, he or she enters the password which was set originally. The entered password is then compared with the stored hashes. If both of these match, then the user is authenticated to carry on to the Windows account.

Whenever a hacker tries to hack Windows, he needs to get these hashes cracked. Both manual and automated methods can be used by the hacker in gaining access to the password. As opposed to email hacking, this may not require password access. Windows hacking needs to be proceeded through password access. The most prevalent methods which are used to crack the passwords of a Windows account include:

- ✓ Brute-force method
- ✓ Rainbow table attack

Types of Windows Hacking Attacks

- To monitor and control all types of accounts operated on a computer.
- To make alterations in the original password without the knowledge of the user.
- To create a new user account.
- To delete the already existing account of the user.
- To create an invisible account in a computer system.

Barriers for a Windows Hacking Attack

- Using the BIOS setup; a change in the boot sequence can be a good measure. The sequence of first, second, and third boot device can be as follows:
  - Hard Disk (1st boot drive)
  - DVD/CD drive (2nd boot device)
  - Removable (3rd boot device)

- Creating a stronger BIOS password
- Implementing a physical lock of your personal computer cabinet

Chapter 4: Hacking Web Servers

A web server can be categorized as the program which uses HTTP (Hyper Text Transfer Protocol). The web server configures the different kind of web pages. Web servers deal with various kinds of HTML documents including the images, objects, scripts, and text.

The web server comprises of an IP address and a specific domain name. For instance, if you type the URLhttp://www.beginnerguide.co./mobile.html in the browser tab, it will enter an appeal to the server which exists with the domain name beginnerguide.co. The server will then process this application and will get access to this page. The browser will send this page to the user who requested this URL.

Any computer system can be converted into a working web server. It will require installing various server software. After installation, the respective machine will be connected to the Internet.

A number of different applications for the web server are available in the web server market.

In order to setup a web server, the software which is available may include:

- ✓ IIS
- ✓ Apache

The Workings of a Web Server

While using your personal computer, laptop, or any other electronic gadget, you can surf millions of websites all over the world. All you need is to type the URL in a browser and enter it. No matter which URL you request, it will appear on the screen. The geographical or virtual location does not matter in this case.

In this phase, the web browser has actually formed a connection with the web server, requests the specified page, and makes it available for the end user.

Login Process for Websites

Different websites allow different accounts in order to login for a particular account. When you type these URL in the task bar, the server will direct you to the website.

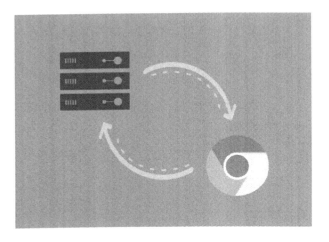

To sign in to the account, the user specifies the password, username, and other details. These ultimately allow the user to get access to the account.

The web server responds to these passwords and user names and sends those to the major database server.
The database server then senses these passwords and usernames and carries an analysis to authenticate the access. The different kinds of table checking enable the access to the server account.

Web servers use authentication results which are provided by the database server. The eventual results of authentication will send the user to the requested page.

If the authentication is complete, the sign-in process will be initiated, otherwise the user will be requested to provide the information again.

Types of Hacking Attacks on Web Servers

There are multiple types of web server attacks which can extend to these servers by many different types of hackers. Some of these include:

> **Web Ripping**

Web ripping deals with mediafiles and pictures. When a hacker web rips, he finds these files and pictures on certain URL and extracts these in a particular way. These pictures are then saved in the hard drive of the hacker's computer.

Web ripping also enables an in-depth copying of the website profile. This profile is then shifted to the local disk. It enables malicious access to the links and fields of the website.

In the case of web ripping, the most efficient tool that is being used is the Black Windows web ripper.

> **SQL Injection**

The hackers which use SQL injection are usually aware of the vulnerabilities which are a part of the web server. The hackers exploit these vulnerabilities, which eventually lead to the permission and access to the database. The database can then be read, altered, or modified.

An example of a SQL injection hacking attack involves enabling a true condition by feeding the similar value into the web page. These values are fed into the system of login as mentioned below:

✓ Login: 2' or '2'='2 and Password= 2' or '2'='2

✓ Login: 2' or '2'='2';--

When the argument for the username is evaluated, '2'='2' will come out to be TRUE. In this case, the returned value will be the authentic username.

➢ PHP Remote Code Execution

This hacking attack is based on the programming language PHP. The hacker, in this case, carries out the code at the personal system level which is then forwarded to the targeted web server. With this step, the attacker can circumvent the web server. Bypassing allows the hacker to contact and gain the files with the full and special rights which are cherished by the original server system software.
A multiple number of PHP programs enclose a susceptibility that can send the inauthentic users to the web server. The user will be directed unchecked and unnoticed, so any hacker can use this PHP program.

➢ Google Hacking

Among the list of search engines, Google is the most used one.

Google keeps a record of the pages crawled by users by taking snapshots of pages. It will access these through the cached link which are shown on search result pages.

Google hacking makes use of advance search operators through the Google search engine; it will locate the precise text strings between the search results. Some hackers may find some appropriate versions of susceptible web applications.

The hacker can target particular password files, directories, file types, or different cameras with specific IP addresses.

➢ Directory Transversal Attacks

Directory traversal hacking attacks enable the malicious users to specifically "navigate" the directory and sidestep the access control list to gain access to the hidden files. These types of hidden files can enable the hackers to manipulate the data and use it for destructive purposes.

These attacks are based on HTTP exploits. These attacks commence with an easy GET or various types of requests based on HTTP. Many web servers can be victim of this type of attack, which can encounter vulnerabilities very easily in a much shorter duration, as compared to other types of hacking attacks.

Conclusion

The world of technology and information has travelled a long way through development and progress. It is through this development that the whole of mankind can cherish the fruitful effects of easy communication, access, and collaboration. The world is truly a global village. But as the development and progress is aggravated, a number of new challenges have occurred which demand the human intellect to devise useful ways of handling these challenges.

One such challenge which has gained desperate attention during last few decades pertains to the activity of hacking. Hackers have been involved in extending a number of extensive losses to different computing systems and networks, both in terms of financial and non-financial quantum. So all users and developers of computer systems are very much involved in knowing the science and art of hacking.

People are now interested in gaining knowledge about hacking so that they no longer remain vulnerable to hacking activities and subsequent losses. This book has been written with an intention of giving knowledge to even the most basic users of computers and networks because the victim is the most crucial person who needs to know about the activity. I have outlined the most basic knowledge so that you can start from the rawest outline of knowledge pertaining to hacking. I wish and hope that the information presented has been useful and understandable for my readers.

Thank you for reading. I hope you enjoy it. Ask you to leave your honest feedback.

Made in the USA
Middletown, DE
09 March 2017